SCIENTIFIC AMERICAN EDUCATIONAL PUBLISHING

LET'S CELEBRATE!

10 FUN EXPERIMENTS FOR THE HOLIDAYS

BRING SCIENCE HOME

Published in 2024 by The Rosen Publishing Group, Inc.
2544 Clinton Street, Buffalo, NY 14224

Contains material from Scientific American , a division of Springer Nature America, Inc., reprinted by permission as well as original material from The Rosen Publishing Group.

First Edition

Editor: Kristen Rajczak Nelson
Designer: Rachel Rising

All illustrations by Continuum Content Solutions

Activity on p. 5 by Science Buddies, December 26, 2013; p. 11 by Science Buddies, February 14, 2013; p. 17 by Science Buddies on March 13, 2014; p. 23 by Science Buddies on July 10, 2014; p. 29 by Science Buddies on July 4, 2013; p. 35 by Daisy Yuhas on October 25, 2012; p. 39 by Science Buddies on October 30, 2014; p. 45 by Science Buddies on October 24, 2013; p. 49 by Science Buddies on November 27, 2014; p. 55 by Science Buddies on November 21, 2013.

Photo Credits: pp. 5, 11, 17, 23, 29, 35, 39, 45, 49, 55 Anna Frajtova/Shutterstock.com; pp. 5, 8, 14, 23, 26, 29, 33, 39, 42, 49, 52 cve iv/Shutterstock.com.

Cataloging-in-Publication Data
Names: Scientific American, Inc.
Title: Let's celebrate!: 10 fun experiments for the holidays / edited by the Scientific American Editors.
Description: New York : Scientific American Educational Publishing, an imprint of Rosen Publishing, 2024. | Series: Bring science home | Includes glossary and index.
Identifiers: ISBN 9781725349940 (pbk.) | ISBN 9781725349957 (library bound) | ISBN 9781725349964 (ebook)
Subjects: LCSH: Holidays--Juvenile literature. | Festivals--Juvenile literature. | Science--Experiments--Juvenile literature. | Cookbooks--Juvenile literature.
Classification: LCC GT3933.L48 2024 | DDC 394.26--dc23

Manufactured in the United States of America

CPSIA Compliance Information: Batch #CWSA24. For further information contact Rosen Publishing, New York, New York at 1-800-237-9932.

CONTENTS

Introduction 4

Scrumptious Science: How Does Baking Powder Make Cornbread Fluffy? ⚛ 5

Stethoscope Science: Hearing Heart Rates ⚛ 11

Surfactant Science: Make a Milk Rainbow 17

Saucy Science: Exploring the Science of Marinades ⚛ 23

Spectacle Science: Exploring Homemade Rockets ⚛ 29

Spooky Science: Make a Ghostly Illusion 35

The Sounds of Halloween ⚛ 39

Discovering the Eerie Colors Behind Afterimages 45

Great Globs of Gluten! ⚛ 49

Salty Science: Is There Iodine in Your Salt? 55

The Scientific Method 61

Glossary 62

Additional Resources 63

Index 64

⚛ THESE ACTIVITIES INCLUDE SCIENCE FAIR PROJECT IDEAS.

INTRODUCTION

Dazzling fireworks exploding in the sky on the Fourth of July and the exciting chills of watching your favorite scary Halloween movie—science is behind these and many more fun U.S. holiday traditions! The experiments in this book will help you understand this science. You'll also discover ways to liven up celebrations, including learning the secrets of many popular holiday recipes.

Projects marked with include a section called Science Fair Project Ideas. These ideas can help you develop your own original science fair project. Science fair judges tend to reward creative thought and imagination, and it helps if you are really interested in your project. You will also need to follow the scientific method. See page 61 for more information about that.

Scrumptious Science

How Does Baking Powder Make Cornbread Fluffy?

TRY OUT THIS TASTY TASK TO LEARN THE CHEMISTRY BEHIND THE SOUTHERN NEW YEAR'S DAY TRADITION OF MAKING CORNBREAD.

Do you ever wonder about the purpose of each of the ingredients in your favorite recipes? For example, why is baking powder used in some muffin recipes? How does the baking powder affect how the muffins look, feel, and taste? In this science activity, you will use a scientific method to make some cornbread muffins and find out. And then you can decide on the best recipe to use if you want to make cornbread for New Year's Day, which is part of a common tradition in the South.

PROJECT TIME

1 hour, plus time to preheat oven

KEY CONCEPTS

Chemistry
Food science
Recipes

BACKGROUND

If you examine the texture of a muffin or a slice of bread, you'll see that the basic structure is a random sort of meshwork that surrounds pockets of air. Without these air spaces, the end result would more closely resemble a brick than bread. The air spaces are created by bubbles of gas (mostly carbon dioxide) produced within the batter during the baking process. Ingredients that produce these bubbles are called leavening agents.

In "quick breads," which have no yeast (and don't require hours for dough to rise), the leavening agent is usually baking powder. Baking powder is typically a mixture of corn starch (which helps keep ingredients dry), sodium bicarbonate (a base, also known as baking soda), sodium aluminum sulfate (an acid) and monocalcium phosphate (another acid). When baking powder dissolves in liquid ingredients, the sodium bicarbonate reacts with the sodium aluminum sulfate and monocalcium phosphate. A product of this chemical reaction is carbon dioxide gas, the leavening for the baked product. As the batter bakes, the carbon dioxide produced by the baking powder makes bubbles in the batter. The bubbles are trapped by the surrounding structure of the batter, mainly supported by proteins in the flour and eggs.

MATERIALS

- Muffin pan that holds at least six muffins
- Paper muffin cups for lining the pan (optional)
- Oven
- Three mixing bowls or other large bowls
- Measuring cups and spoons
- 3/4 cup of all-purpose flour
- 3/4 cup of cornmeal
- 3/4 tablespoon of sugar
- 3 teaspoons of baking powder
- 3 tablespoons of butter
- Two eggs
- Two small bowls
- 3/4 cup (177 ml) of milk
- Fork
- Oven mitts
- Cooling rack
- Timer
- Volunteers to taste test! (optional)

PREPARATION

- Butter the muffin pan or line it with paper muffin cups.
- Preheat the oven to 425°F (218°C). Be sure an adult is present to help when using the oven during this activity.
- Prepare three mixing bowls, or other large bowls, with the dry ingredients for making three different batches of muffins. In each bowl, mix together 1/4 cup of all-purpose flour, 1/4 cup of cornmeal, and 1/4 tablespoon of sugar. (Do not add any baking powder yet.)
- Melt 3 tablespoons of butter in a small bowl.
- Lightly beat two eggs in a separate small bowl.

PROCEDURE

- Take one of the mixing bowls you prepared with the dry ingredients. You will add no baking powder to this batter—just go ahead and make a well, or hole, in the center of your dry ingredients, and mix in half of a slightly beaten egg, 1/4 cup (59 ml) of milk, and 1 tablespoon of melted butter. Mix the batter until it is smooth. *How do you think the muffins will turn out when baking powder is not added?*
- Fill three of the (buttered or paper-lined) muffin cups in the pan about two-thirds full with the batter. Do your best to make each muffin with the same amount of batter, but do not spend too much time on this.
- Take one of the other mixing bowls you prepared with the dry ingredients and add 1 teaspoon of baking powder to it. This is the amount of baking powder recommended for this recipe. Similarly, mix in half of a slightly beaten egg, 1/4 cup (59 ml) of milk, and 1 tablespoon of melted butter. Mix the batter until it is smooth. *Why do you think it's important to bake the muffins shortly after mixing the baking powder with wet ingredients?*

- Quickly fill three of the remaining muffin cups in the pan about two-thirds full with this batter, again trying to fill them similarly but without spending too much time on this because the baking powder starts working as soon as it touches liquid.
- Bake at 425°F (218°C) for 15 to 20 minutes, or until the muffins are nicely browned and baked through. When the muffins are done, place the pan on a cooling rack for a few minutes, and then tip the muffins out. Set them aside for now.
- Take the last mixing bowl you prepared with the dry ingredients and add 2 teaspoons of baking powder to it. *How do you think the muffins will turn out when the batter has extra baking powder?* Then similarly mix in half of a slightly beaten egg, 1/4 cup (59 ml) of milk, and 1 tablespoon melted butter. Mix the batter until it is smooth. Fill three of the (newly buttered or paper-lined) muffin cups in the pan about two-thirds full with the batter, distributing it equally. Bake the muffins as you did before, and then let the pan cool on a cooling rack again.
- When all three muffin batches are done, compare how they each look, feel, and taste. *How do the muffins baked without baking powder compare to the other muffins? How about the muffins baked with extra baking powder? How are the muffins similar and how are they different? Can you explain this in terms of what baking powder does?*

EXTRA

Try repeating this activity but use different amounts of baking powder. You could try 1/2 teaspoon or 3 teaspoons. *How does using other amounts of baking powder affect how the muffins turn out?*

SCIENCE FAIR IDEA

If you have a kitchen scale, you could measure the weight of each muffin, and even calculate their densities by measuring the dimensions of the muffins. *How do the weights of the muffins compare to each other? What about their densities? Do your results make sense to you?*

SCIENCE FAIR IDEA

Pick a different ingredient from the muffin recipe. Do research to understand the function of the ingredient in the recipe, and then predict what will happen when you change the proportion of that ingredient in the recipe. *Can you accurately predict how changing the amount of a different ingredient changes how the muffins turn out?*

OBSERVATIONS AND RESULTS

Did the muffins baked without baking powder turn out shorter and harder than the muffins from the other batches? Were the muffins made with 1 or 2 teaspoons of baking powder much more similar to each other?

Baking powder is a leavening agent that produces carbon dioxide gas during the baking process. The carbon dioxide gas bubbles become trapped in the batter as it bakes, forming air spaces in the resultant muffins. This is why the muffins baked without baking powder should have had many fewer air spaces, which made them flatter and denser than the other muffins. These muffins may have tasted and felt more like bread than fluffy corn muffins. The muffins baked with 1 teaspoon of baking powder should have seemed like typical corn muffins—they should have been somewhat fluffy with many air spaces. The muffins baked with 2 teaspoons of baking powder may have seemed very similar to the muffins with 1 teaspoon of baking powder, indicating that additional baking powder did not make a huge difference in how the muffins turned out. That said, the muffins baked with 2 teaspoons of baking powder may have seemed slightly fluffier than the ones baked with 1 teaspoon of baking powder, but overall, these muffins should have been much more similar to each other than the ones made with no baking powder.

CLEANUP

Enjoy some tasty corn muffins! And be sure to clean and put away the all materials you used to make them.

Stethoscope Science

Hearing Heart Rates

THIS VALENTINE'S DAY, CELEBRATE THE HEARTS THAT BEAT INSIDE ALL PEOPLE BY CREATING A HOMEMADE STETHOSCOPE TO MEASURE HEART RATE!

Around Valentine's Day, hearts seem to be everywhere. But our actual hearts are at work all year—every hour, minute, and second. The heart works hardest when we physically exert ourselves. How does its beat change? A doctor can figure this out by using a tool called a stethoscope, which is a long, thin plastic tube that has a small disc at one end and earpieces at the other. In this activity, you will make a homemade stethoscope and use it to measure peoples' heart rates at rest and after exercising.

PROJECT TIME

15 to 20 minutes

KEY CONCEPTS

Stethoscope
Health
Exercise
Amplification
The heart

BACKGROUND

You're probably familiar with how a stethoscope is used from visits to your own doctor. To listen to the heart, the doctor puts the stethoscope's flat disc or hollow cup on a patient's body and the earpieces go into the doctor's ears. But how does the stethoscope work? How does it allow the doctor to hear sounds inside of the patient's body?

The disc and the tube of the stethoscope amplify small sounds such as the sound of a patient's lungs, heart, and other sounds inside the body, making them louder. The amplified sounds travel up the stethoscope's tube to the earpieces that the doctor listens through. Heartbeats can easily be heard using a good stethoscope. Every time a person's heart beats it contracts and acts as a powerful pump, which circulates blood that carries oxygen and nutrients throughout the body.

MATERIALS

- Duct tape or other strong tape
- Scissors
- Plastic funnel
- A cardboard tube from a paper towel roll
- A volunteer who can safely exercise vigorously for one minute
- Stopwatch or clock that counts seconds

PREPARATION

- Make sure you have a volunteer who can safely exercise vigorously for one minute.
- Put the narrow end of the funnel into the cardboard tube.
- Using a strip of duct tape or other strong tape, tape the funnel and cardboard tube together. Make sure there are no gaps or spaces where you tape them together.

- Your stethoscope is now ready to use! Practice listening to the heartbeat of a volunteer by putting the funnel on the left side of the volunteer's chest. Make sure the funnel is flat against their chest. Put your ear against the hole at the end of the cardboard tube. *Do you hear a heartbeat?*
- Before you begin the activity, have your volunteer sit quietly in a chair for at least five minutes.
- **Tip:** If it's noisy or the volunteer is wearing thick clothing, it may be hard to hear the heartbeat, so you may need to adjust conditions accordingly.

PROCEDURE

- After the volunteer has been resting in a chair, listen to their heartbeat and count how many times it beats in 15 seconds.
- Multiply this number by four. This is the resting heart rate of the volunteer in beats per minute (bpm). *What is the volunteer's resting heart rate?*
- Ask the volunteer to exercise in place for one minute by doing jumping jacks or running in place. Right after the volunteer has stopped exercising, listen to their heartbeat and count how many times it beats in 10 seconds. *Why do you think you do this for only 10 seconds instead of 15?*
- Multiply this number by six. This is the heart rate right after exercising in bpm. *What is the volunteer's heart rate now?*
- *How did the heart rate change after exercise? Why do you think it changed like it did?*
- *If a person regularly exercised, how do you think this would change their heart rate? How do you think that person's heart rate during rest and exercise would be different?* **Hint:** Think about how regular exercise may change the heart.

EXTRA

In this activity, you only tested one person's heart rate, but you could try your stethoscope on several different people. *How do different peoples' heart rates compare with one another? Do they all change a lot after exercising or do some change only a little bit or not at all? Does the heart rate correlate with a factor such as age, biological sex, or body mass index (BMI)?*

EXTRA

A person should use at least 50 percent of their maximum heart rate when they are exercising for the activity to qualify as exercise. You can find out more about maximum and target heart rates and then apply this information to create an exercise routine. *Based on this information, what physical activities qualify as exercise?*

SCIENCE FAIR IDEA

There are several different ways you can make a homemade stethoscope. For example, instead of a cardboard tube you could use a short piece of garden hose or plastic tubing. Or you could try varying the tube's length. You could even try different size funnels. *What homemade stethoscope design works best? What goes into making a good stethoscope design?*

OBSERVATIONS AND RESULTS

Could you hear the volunteer's heartbeat using the homemade stethoscope? Did you find that when they exercised their heart rate increased compared with what their heart rate was when they were resting?

When people exercise, their bodies need more oxygen, and consequently their hearts beat faster, and their heart rates increase. This is why you most likely found that when your volunteer exercised, their heart rate increased compared with their resting rate. In addition, genetics, biological sex, age, and health all affect people's heart rates. The rates in people who exercise regularly usually will not increase as much during exercise and will return to a resting rate more quickly after exercise is stopped. Regular exercise strengthens the heart so that it does not need to work as hard.

Although you can determine someone's resting heart rate by counting the number of beats in 15 seconds and multiplying by four to get the beats per minute, to calculate a heart rate immediately after exercise, it is better to count the number of beats for 10 seconds and multiply that value by six (to get the beats per minute). Because the heart will quickly slow down after exercise ceases, its rate should be measured immediately after a person has stopped exercising (or while they exercise, if possible).

CLEANUP

Put the materials to make the stethoscope back where you found them and save your stethoscope to measure more heart rates!

Surfactant Science

Make a Milk Rainbow

MAKE A RAINBOW FIT FOR A POT OF GOLD USING MILK, LIQUID SOAP OR DETERGENT, AND FOOD COLORING WITH THIS FUN ST. PATRICK'S DAY ACTIVITY!

Have you ever made decorative rainbows for Saint Patrick's Day? They can be fun to make using colorful construction paper or other craft supplies. But did you know you can make a simple one with milk, liquid soap or detergent, and food coloring? How the rainbow is created by this mixture might surprise you! In this science activity, you'll make your own milk rainbow and explore how detergent and surface tension are involved in its creation.

PROJECT TIME

10 to 20 minutes

KEY CONCEPTS

Physics
Chemistry
Liquids
Molecules
Surface tension
Soap

BACKGROUND

In a liquid, the molecules (small individual particles that can have positive and negative charges on their surfaces) can, just like magnets, attract and repel one another. In the case of a water molecule surrounded on all sides by other water molecules, there is roughly the same amount of pulling and pushing force. The overall effect is no change. But at the surface, where the water molecules are exposed to air on one side, the water molecules experience more pulls downward, toward the other water molecules below them, than upward toward the air. This causes the water at the surface to contract, minimizing its surface area. This phenomenon is called surface tension.

Certain compounds affect water's surface tension: Some are hydrophilic (or polar), which are attracted to water. Others, however, are hydrophobic (or nonpolar) and repel water. Some compounds have both properties. When such a compound is added to water, the hydrophilic end will try to get close to the water molecules whereas the hydrophobic end will push away from them. This pulling and pushing separates the water molecules from one another and consequently decreases surface tension. Compounds that lower water's surface tension are known as surfactants.

MATERIALS

- Milk; 1 percent, 2 percent, or whole
- Plate or some other type of shallow dish, such as an aluminum pie tin (Make sure the bottom is flat and not wrinkled.)
- Red, yellow, green, and blue food coloring. Other or fewer colors could also be used to make something other than a four-color rainbow.
- Cotton swab
- Liquid dish soap or liquid laundry detergent
- Paper towels (useful for cleaning up afterward)

PREPARATION

- Because this activity involves dyed milk, it is best to do the activity outdoors or on a surface that can safely have dye spilled on it, in case an accident occurs. Doing this activity near a sink or hose makes it easier to clean up at the end.

PROCEDURE

- Slowly pour enough milk into the plate so that the bottom is completely covered. Allow the milk to settle for a moment.
- Add one drop of each color of food coloring to the milk, forming a horizontal line of drops near the bottom of the plate. To make a simple, four-color rainbow, add the colors in the following order, going from left to right: red, yellow, green, then blue. *How do you think the drops will become a rainbow?*
- Touch one end of the cotton swab right below the middle of the line of drops, on the plain milk. *What happens?*
- Take the other end of the cotton swab and rub some liquid dish soap on it. Touch the soap-covered end to the same area as before, right below the middle of the line of drops, on the plain milk. *What happens this time?*
- Try touching other areas of the milk with the soap-covered cotton swab end. *What happens as you continue to touch the milk in other places?*
- Hold the soap-covered cotton swab end in one spot for a few seconds. *What happens? If you hold the swab in the milk long enough, do you no longer see the same effect?*

EXTRA

Try this activity with milks that have different amounts of fat (for example, nonfat, 1 percent, 2 percent, whole, etcetera). *How do the results vary when using different types of milk?*

EXTRA

You should have seen a simple, four-color rainbow in this activity, but you can also try to make a more complex, seven-color rainbow. To do this you will need to mix together some food coloring to make orange, indigo, and violet, and then add them to the line of drops on the plate. For example, you could try mixing one drop of red and one drop of yellow food coloring to make orange. *Can you make a seven-color rainbow this way?*

EXTRA

This activity used milk, but you could try other liquids, too, such as water, clear soda, and melted butter. *Did you get the same result using liquids other than milk?*

OBSERVATIONS AND RESULTS

Did you see the food coloring move away from the soap-covered cotton swab when you touched it to the milk, making a four-color rainbow streak across the plate?

When you touched the plain cotton swab (without soap) to the milk, you should not have seen any effect. When you touched the soap-covered end to the milk, however, you should have seen the food coloring quickly move away from the cotton swab. Because a line of food coloring drops was made in the milk, putting the cotton swab below the middle of the line should have made the food coloring form a four-color rainbow as it all streaked and swirled up the plate, moving away from the cotton swab.

Detergent, such as liquid dish soap, is mostly surfactants. These can lower the surface tension of water as well as milk, which contains water and molecules of fat. Surfactants have a hydrophilic part that wants to interact with the water and a hydrophobic part that wants to interact with the fat molecules. Because of this, when the cotton swab with soap touched the milk, the soap separated the fat from the water in the milk, dissolving the fat (which is how soap cleans greasy, dirty dishes). This also decreased the milk's surface tension. As the soap spread out from the cotton swab, it decreased the milk's surface tension around it, and the higher surface tension surrounding this area pulled the milk (along with its food coloring) toward it. If enough soap is added, however, the soap and milk become evenly mixed and the milk (and food coloring) no longer move when more soap is added.

CLEANUP

Do not drink the dyed milk because it has soap in it. Instead, carefully pour the dyed milk directly down the sink drain, trying not to splash the sink itself. (You should also run water as you pour.) If you need to move the plate to a sink to do this, it is easiest to add a few paper towels to the plate to soak up some of the dyed milk (and help prevent spills) before attempting to move the plate.

Saucy Science

Exploring the Science of Marinades

LEARN THE SECRETS BEHIND WHAT MAKES DELICIOUS MARINADES STICK TO FOODS TO LIVEN UP YOUR NEXT FOURTH OF JULY BARBECUE!

PROJECT TIME

About 1 hour

KEY CONCEPTS

Food science
Cooking
Marinades
Barbecue
Acids

Have you ever tasted delicious grilled chicken and wondered how it got so much flavor? Maybe you have heard your family talk about marinating foods before cooking or grilling them. A marinade is a mixture of seasonings used to flavor or tenderize food. Some cooks have strong opinions about the best way to marinate their favorite food, be it a steak, tofu steak, chicken breast, or veggie kebobs. One of the keys to a good marinade is that it stays on the food you are planning to cook. In this activity, you will do a test to see what factors might be most important in making a marinade stick to the surface of food. After doing this activity you could use your findings to help you make tastier food for a summer BBQ!

BACKGROUND

Every culture has its own unique way of preparing food. But whether a recipe is Chinese or American, Italian or Indian, some of the main dishes usually call for a marinade. Some of the original marinades from several centuries ago were briny (very salty) liquids, such as seawater, which helped preserve foods before refrigeration was available. Whatever they are made of, marinades are usually meant to preserve, tenderize, and flavor foods.

In this activity, you'll test how various ingredients affect the adsorption (yes, with a "d") of a marinade ingredient onto the surface of a food. (The word "adsorb" is used to describe the process by which a substance adheres to the surface of an object, as opposed to being absorbed into it.) The ingredients you'll test are salt, vinegar, and sugar. Salt obviously makes foods salty and sugar makes them sweet, but how well do these ingredients work for making a marinade be adsorbed to food? Vinegar is an acid—other acids you might find around the kitchen include lemon juice and orange juice. Does something acidic make a good marinade? Along with these ingredients, instead of actual seasoned marinades, you'll use a food dye so you can see the level of adsorption on the food.

MATERIALS

- Knife
- Cutting board or dinner plate
- Tofu, extra-firm (You can use meat if you choose, but if so, use an adult helper and be sure to wash any surfaces—including your hands—that the raw meat comes into contact with. The dye works well with chicken breast.)
- Four plastic cups, each at least 9 ounces (266 ml) in size
- Masking tape or painter's tape and a pen or permanent marker (This is so you can label the cups. If you are using disposable cups, you can just use a permanent marker.)
- Food dye (A dark color, such as blue or green, should be used.)
- Salt
- Distilled white vinegar
- Sugar
- Tap water
- Clean spoons or other utensils for stirring

- Measuring cups
- Measuring tablespoon
- Sheet of white paper
- Clear plastic wrap
- A fork or a slotted spoon
- A timer or clock

PREPARATION

- Cut the tofu block into eight cubes, where each cube is about 0.5 inch (1.3 cm) on each side.
- Label four plastic cups as follows: "water," "salt," "vinegar," and "sugar."

PROCEDURE

- To the cup labeled "water," add 1 cup (237 ml) of water. To the cup labeled "salt," add 1 tablespoon of salt and 1 cup (237 ml) of water. To the cup labeled "vinegar," add 1/4 cup (59 ml) of vinegar and 3/4 cup (177 ml) of water. To the cup labeled "sugar," add 1 tablespoon of sugar and 1 cup (237 ml) of water. *How well do you think the different ingredients will stick to the tofu cube's surface?*
- Add eight drops of food dye to each of the cups.
- Using a clean spoon or other utensil for each cup, stir the solutions so that the dye gets mixed in and the salt and sugar are completely dissolved in their cups.
- Carefully add two tofu cubes to each of the four cups. Let the tofu cubes marinate in the cups for 45 minutes at room temperature. (Do not worry if the cubes in the salty water float.)
- While the cubes are marinating, lay a white sheet of paper on a nearby surface and cover it with clear plastic wrap.

- When the cubes are done marinating, use a fork or slotted spoon to remove the cubes from each cup. Place the cubes on the sheet of paper, leaving the cups that held the cubes nearby so you know which cup each cube came from.
- Examine the cubes. (If not all of the sides on a cube are equally dyed, examine an average-looking side on the cube.) *Which cubes became the most dyed? Which are the least dyed? How do they compare with the cubes from the cup that only had water and food dye? What does this tell you about how well the different ingredients (sugar, salt, and vinegar) stick to the surface of the food?*

SCIENCE FAIR IDEA

There are many variables that you could try changing in this activity, such as what type of food you test, the amount of time the tofu cubes are marinated, the temperature they're marinated at, and the concentration of the ingredients tested. (If you use chicken breast or other meat, be sure to wash everything that comes into contact with the raw meat—for instance, raw chicken might carry salmonella bacteria.) When exploring other variables, be sure to only change one at a time, keeping everything else the same. *How well do the tofu cubes become dyed when marinated for a longer amount of time, at a different temperature, or with a different concentration of ingredients?*

EXTRA

Some marinades have vegetable or olive oil in them. *What happens if you mix oil in with the marinade?*

EXTRA

You could cut the tofu cubes with a knife after they are done marinating to look at the inside of the cubes. *How far has the dye been absorbed into the cubes from the different cups?*

OBSERVATIONS AND RESULTS

Did the tofu cubes in the cup with the vinegar become the most dyed? Were the other tofu cubes less dyed, but all dyed similarly to each other?

You should have seen the tofu cubes that were marinated in the cup with the vinegar/food dye/water solution clearly become the most dyed—or had adsorbed the dye best—compared with any of the other tofu cubes. Vinegar is an acid, like lemon juice and orange juice. Acids can cause foods such as meats to be broken down, or tenderized. (You may have heard people warn not to use too much vinegar in a marinade. This is because the acid can tenderize the food so much that it becomes mushy!) When vinegar is used in a marinade, it breaks down the food's surface and lets the marinade be adsorbed there. Salt works well in marinades for meat, too, because it helps break open the cells, allowing the marinade to penetrate into the tissue. Sugar in marinades helps the food taste sweet, but does not help marinades be adsorbed into food as much.

CLEANUP

It is not recommended to eat the tofu cubes used in this activity. You could compost them or throw them in the trash. If you used chicken breast or other meat, discard it in the garbage and be sure to wash everything that came in contact with the raw meat, especially if you used raw chicken, which might have salmonella.

Spectacle Science

Exploring Homemade Rockets

THREE . . . TWO . . . ONE . . . BLASTOFF! USE TWO SIMPLE HOUSEHOLD INGREDIENTS TO LAUNCH A HOMEMADE "ROCKET" AND GIVE YOUR FRIENDS OR FAMILY A SHOW THIS FOURTH OF JULY!

PROJECT TIME

1 hour

KEY CONCEPTS

Chemical reactions
Physics
Gravity
Rockets
Thrust
Pressurization

Have you ever marveled at how fireworks, toy rockets, or real spacecraft can launch into the air? It can be an amazing thing to witness. It is thrilling to see something lift off against Earth's gravity. The strong push required to launch a spacecraft comes from a chemical reaction in its rockets. This means that every time you see a spacecraft launch, you're watching chemistry at work. In this activity, you'll get to blast an object into the air using two simple household ingredients: baking soda and vinegar. Investigate how to mix these chemicals to get the best lift off, and then this Independence Day you could give your family a homemade, gravity-defying show!

BACKGROUND

How does a spacecraft lift off and get into space? The simple answer is that it has rocket engines that propel it. The rockets depend on combustion to provide the thrust the spacecraft needs to overcome the force of gravity and climb into orbit. Combustion is a fast, exothermic chemical reaction between a fuel (for example, jet fuel) and an oxidizer (such as oxygen) in which the fuel burns and heat is produced. Usually, the fuel is an organic compound (containing hydrogen and carbon, and sometimes metal and/or other components). During the chemical reaction, new compounds are made. These are referred to as the exhaust. The rockets push the hot exhaust out from the bottom at high pressure and thus the spacecraft is thrust upward.

In this activity, instead of using rocket fuel, you will use baking soda (sodium bicarbonate) and vinegar (acetic acid) to make a different kind of chemical reaction that can launch a small-scale rocket made from a film canister. The reaction produces water and carbon dioxide (which will appear as bubbles). You'll take advantage of the pressure the carbon dioxide gas makes in the capped film canister to launch your rocket.

MATERIALS

- Plastic film canister with a lid and tight seal. Fuji or Kodak canisters should work.
- Baking soda
- Measuring spoons
- Wax paper or bowl
- Spoon
- Water
- Vinegar
- An open outdoor area at least 6.5 feet (2 m) from buildings. It is ideal to have a hard, flat surface such as a paved patio or driveway.
- Safety goggles
- Rag or paper towel
- Optional: Construction paper, transparent tape, stickers, and scissors
- Optional: A helper to watch, a helper to take a video, or a video camera with a tripod

PREPARATION

- If you like, you may decorate your film canister rocket. You could wrap a piece of construction paper around the canister and cut the paper so it just covers the rocket's sides (but does not go above or below the sides). After evenly wrapping the paper on the canister, secure it with some tape. You can add additional flat decorations, like stickers or drawings. Make sure it is still easy to put the lid on.
- Remember, when you launch your film canister rocket, be sure to wear eye protection and exercise caution!

PROCEDURE

- Place 1 teaspoon of 31 onto the wax paper or bowl. Add 1/8 teaspoon (0.5 ml) of water to the baking soda and mix it in well. If you're using wax paper, you can carefully use the wax paper to fold the damp baking soda onto itself to help mix in the water.
- Turn the film canister lid upside down and pack the inside of the depression with the damp baking soda. (Do not put baking soda near the rim where the canister snaps onto the lid.) Pack it tightly. Turn the lid right side up again for a moment. *Does the damp baking soda stay in place?* If it stays, move on to preparing the vinegar. If it falls out, add a little bit more water to the baking soda and mix it in, but try to add as little water as needed. The baking soda will not need to stay packed into the lid long.
- Add 1 teaspoon (5 ml) of vinegar to the canister at a time, filling it almost to the top. You need to add as much vinegar to the canister as possible—just enough to keep the vinegar and the baking soda from coming into contact when you later snap the lid onto the canister. Depending on the canister, this may be about 5 teaspoons (25 ml) of vinegar. *How much vinegar did you use?*

- Go outside to an open area at least 6.5 feet (2 m) from buildings. If you want to videotape the reactions, set the video camera so that it has in its viewfinder on the spot where you will launch your canister rocket and the equivalent of at least the first story of a building and then start the video. (Alternatively, you may have a helper watch the reactions to help you figure out how high the canisters go.)

- Put on your safety goggles. Stoop down near the ground on a flat, hard spot and quickly snap the lid onto the canister to seal it. Immediately turn the canister over so the lid is on the ground, and quickly move away. Wait for the chemical reaction to occur. *How long does it take to happen? When the lid pops off, the rocket should launch. How high does the canister go?*

- **Tip:** If the rocket did not launch, the lid might not have been sealed tightly enough. (If this happens you may simply see many foamy bubbles coming out of the canister.) The rocket may not have launched right for some other apparent reason (such as not sealing the lid fast enough). If it didn't launch right, try preparing and launching the canister rocket again. You may need a little practice to get used to launching the rocket.

- After the launch, carefully rinse the lid and canister with water and then dry them. If your canister is covered by construction paper, make sure it doesn't get too wet.

- Prepare the damp baking soda and vinegar as before but this time use a little more than half the original amount of vinegar. For example, if you used 5 teaspoons (25 ml) of vinegar, this time use 3 teaspoons (15 ml). (Still use 1 teaspoon of baking soda.)

- Again, go outdoors, put your safety goggles on and launch your newly prepared canister rocket. *Does it take longer, shorter, or about the same amount of time as the first rocket did to launch? Does it go a higher, shorter, or about the same distance?*

- Lastly, rinse the lid and canister with water, dry them, and prepare them as before but this time use 1 teaspoon (5 ml) of vinegar (or around 1/5 of the original amount that you used). Put your safety goggles on, go outside, and launch the canister rocket. *How long does it take to launch compared with the other two launches? How high does the canister go compared with the previous two times?*
- If you're unsure of any of your results, you can try repeating them (using the same amount of baking soda and vinegar).
- *What amount of vinegar led to the highest launch height? Why do you think this is?*

SCIENCE FAIR IDEA

You can try varying the amount of vinegar even more and see how this affects the rocket's launch. (You could also repeat the same conditions you tested to see how consistent your results are.) *How does changing the amount of vinegar in the canister change how it launches?*

SCIENCE FAIR IDEA

You could also try changing the amount of baking soda (keeping the same amount of vinegar) and see how this affects the canister's launch. For example, you could try comparing 1, 3/4, 1/2, and 1/4 teaspoon of baking soda. (Adjust and use just enough water for the baking soda to stick to the depression in the lid.) *How does changing the amount of baking soda in the lid affect the canister's launch?*

EXTRA

Add a cone and fins to your rocket (such as out of construction paper) and launch it again using the best conditions you found. *How does adding these components affect the canister's launch?*

OBSERVATIONS AND RESULTS

Did the launch using the smallest amount of vinegar result in the highest launch height? Did it also take the most time to launch?

When baking soda and vinegar are mixed together, the reaction produces water and carbon dioxide gas. In the capped film canister, the carbon dioxide gas builds up until the pressure of all of the contained gas causes the canister to pop open. The pressurized carbon dioxide then quickly escapes the canister through the open bottom. This is how the chemical reaction provides the thrust the canister needs to launch.

You may have noticed that when the least amount of vinegar was used, it took a little longer to launch than when more vinegar was used. Because there was less vinegar in the canister, there was more space for carbon dioxide gas to fill. It takes longer for more carbon dioxide to be made from the reaction and thereby more is needed to fill this larger space and build up enough pressure to pop the lid open like it did before. Overall, when the least amount of vinegar is used, more carbon dioxide can fill the canister and a higher launch height should be seen (possibly around 15 feet [4.6 m], compared with around 6 feet [1.8 m] when the canister was nearly full of vinegar).

CLEANUP

If you launched your rocket on a concrete surface, spray the surface down with some water after you have completed your launches.

Spooky Science

Make a Ghostly Illusion

THIS EERIE VISUAL TRICK WILL ADD SOME GHOSTLY FUN AND SCIENTIFIC DISCOVERY TO HALLOWEEN FESTIVITIES!

Halloween is a time for sharing ghost stories and watching spooky movies. But have you ever thought about the science behind some of these uncanny experiences? Haunted houses, for example, take advantage of the way your brain uses sensory information. Often, they include dim lighting and confusing sound effects to keep you disoriented and jittery, primed for fright. In this activity, you'll create an optical illusion and learn that your eyes can play some eerie tricks on you—even in broad daylight. You'll also discover this peculiar apparition has a simple scientific explanation.

PROJECT TIME

About 15 minutes

KEY CONCEPTS

Neuroscience
Photoreceptor
Afterimage
Optical illusion

BACKGROUND

Your eye has many different parts, including a light-sensitive retina at the back of the eyeball. The retina collects light signals via special cells called photoreceptors ("photo" comes from the Greek word for light). Your retina's photoreceptors fall into two groups: rods and cones. Rods perceive changes from light to dark and cones receive color signals. Much like the receptors in your skin, taste buds in your mouth, or hair cells in your ear, your eyes' receptors collect visual information about the outside world and send it to the brain to be decoded and interpreted.

MATERIALS

- Several sheets of white paper
- Brightly colored markers

PROCEDURE

- Draw a ghost (or other shape of your choice) on one of the blank sheets of paper on the center of the page using a colored marker. Make sure your outline is bold or fill the image in with color.
- When your drawing is finished, put your picture side by side with a blank sheet.
- Stare at the center of your drawing for 20 seconds without blinking—it's okay to count aloud to 20 or ask a friend to time you.
- After 20 seconds, immediately look at the blank sheet of paper. *What do you see?*
- Take another sheet and draw another shape, this time using a few different colored markers.
- Again, set your drawing beside a blank sheet of paper, stare at the image for 20 seconds then look at the blank sheet. *Now what do you see? Do you notice any unusual colors?*

- Return to your first ghost drawing. Set it beside a blank sheet and this time stare at the center of the image with just your right eye for 20 seconds.
- Keeping your left eye closed, use your right eye to look at the blank sheet. *What do you see?* Now quickly close it and open your left eye to look at the sheet. *What do you see this time?*
- Switch back and forth between eyes. *Do both eyes see the same image?* Then repeat the previous step, this time starting with your left eye.

EXTRA

Repeat this experiment using both eyes and time how long it takes for the ghostly image to fade away. Try it again, but stare for more or less time. *What do you notice about the image if you stare for more time? What happens if you stare for less time?*

OBSERVATIONS AND RESULTS

Did you see a ghost or shape when you looked at the blank sheet of paper? Did you notice that colors inverted to their complement—red became green, blue turned orange, etcetera?

Illusions can be spooky—especially if you have not learned the science behind them. But they also reveal how your eyes—and brain—work together to process visual information. When you use just one eye at a time, you probably noticed that only the eye that had been staring could see the ghostly illusion. You may have also observed that the ghostly image appeared in different colors. Why was that? The image you saw on the blank paper is called an afterimage.

When you stare at an image for a long period of time, the cones in your eye become overstimulated and lose sensitivity. So when you're looking at, for example, a red ghost, your red-sensitive cones become very active and eventually—if you don't blink—exhausted. When you then look at the blank page, your eyes are still receiving many different color signals. But instead of your cones sending equal signals and balancing out to perceive a white page, your tired red cones don't send a signal whereas the blue and green cones do, leaving you with an aqua-green afterimage.

CLEANUP

Recycle the paper and put the markers and any other materials you used away.

The Sounds of Halloween

MUSIC THAT SEEMS MYSTERIOUS AND EERIE IS PART OF HALLOWEEN FUN. LET'S LEARN THE SCIENCE BEHIND IT!

PROJECT TIME

30 minutes to 2 hours

KEY CONCEPTS

Music
Emotion
Musical instruments
Tempo

Do you enjoy watching scary movies at Halloween, or going to haunted houses? In whatever way you might want to have a thrilling experience this time of year, spooky sounds and music will likely play an important part. One of the things that makes a haunted house or a mysterious movie so eerie is the soundtrack that goes along with it—all the noises orchestrated to raise the hair on your arms, make you shiver, and leave you with little doubt that there are things unknown lurking about. In this science activity, you will investigate what goes into the composition of a perfect creepy soundtrack.

BACKGROUND

Music is a big part of a movie or a theatrical experience. You might not notice it immediately, but music adds layers of richness to the visual experience of a movie, and it enhances the storytelling by increasing emotions or tension in a scene. Scary things feel scarier with the right music. For example, by intensifying emotions, the music makes it so that you don't just jump when that hairy spider comes around the corner—you scream!

There are several different aspects of music that can make it feel eerie or convey a different emotion to us. The musical instruments involved can be important here. For example, percussion instruments such as drums or cymbals make a very different sound than string instruments such as violins or cellos, or horns such as trumpets. The pitch of the melody—high-pitched like bird songs or low-pitched and deep-sounding—and whether it is generally rising or falling also affects how we feel about the music. The music's key (such as whether it is minor or major), tempo (or speed), and volume are other musical components that can each convey different messages to the listener.

MATERIALS

- Age-appropriate children's Halloween movie, such as (depending on the age): *The Nightmare Before Christmas*, *Beetlejuice*, *Wallace & Gromit: The Curse of the Were-Rabbit*, *Coraline*, *Pooh's Heffalump Halloween Movie*, *Hocus Pocus*, *Labyrinth*, *Casper*, *The Witches*, *Corpse Bride*, or one of the *Harry Potter* series. Alternatively, instead of a movie you could use creepy classical and/or movie tracks (from a CD or available online), such as Johann Sebastian Bach's "Toccata and Fugue in D Minor," Modest Mussorgsky's "Night on Bald Mountain," Edvard Grieg's "In the Hall of the Mountain King," Wolfgang Amadeus Mozart's "Requiem Mass," or movie soundtracks by Danny Elfman or John Williams.
- Timer or stopwatch for determining the tempo of the music (optional)
- Pencil or pen and paper (optional)

PREPARATION

- Get your Halloween movie ready to watch. If you have time, you could plan on watching the entire movie, or you could find a few favorite scenes to watch. To do this activity, you will want to listen to at least 20 seconds from at least three different eerie or scary scenes in the movie. Alternatively, if you are using soundtracks instead of a movie, get the music ready to play your favorite songs. You will want to listen to at least three songs.

PROCEDURE

- Watch each eerie or scary movie scene that you picked (or listen to each song) while paying close attention to how the music sounds. The following steps will help you analyze the music. You may need to listen to the music a few times for each scene or song—or for the full movie or composition.
- Try to figure out which main musical instruments are being used. *Do you hear mostly percussion instruments such as the banging of drums or clash of cymbals, string instruments like violins or cellos, big brassy trumpets and other horns, or people singing?* Feel free to jot down notes as you go along.
- Focus on the melody. *Is the melody high-pitched like bird songs or low-pitched and deep-sounding?* Are there rising or falling scales? (A rising scale is like the sequence "do-re-me-fa-so-la-ti-do" and a falling scale is the reverse.)
- Try to listen to the key of the music. A minor key might sound sadder or more serious to you, while a major key might sound more cheerful. (An example of a minor-key song is "My Favorite Things" from the movie *The Sound of Music*, while "Twinkle Twinkle Little Star" is an example of a song typically sung in a major key.) *Is the music in a minor or major key?*

- Now count the beats per minute, or tempo, of the music. If you want, you can tap or clap out the beats yourself for 10 seconds and then multiply by six to get beats per minute. A slow tempo may be less than 70 beats per minute, while a fast tempo may be greater than 110 beats per minute. *Does the music have a relatively fast tempo or a slow one? If you are watching a movie, how does the tempo relate to what's happening in the movie?*
- Also pay attention to the volume of the music. *Is it quiet or loud? If you're watching a movie, how does the volume relate to what's happening?*
- *Overall, what do varied examples of eerie or scary music have in common? How are they different?*

EXTRA

Repeat this activity, but this time use several different movies (and at least five scenes from each movie) to investigate many types of emotional scenes. Listen to see whether certain musical features typically accompany specific types of scenes, such as action, happy, or sad scenes. *Do action scenes usually have the same type of tempo? What about happy and sad scenes—how does the music change with the different types of scenes?*

SCIENCE FAIR IDEA

Get some volunteers and play them short segments of music from different types of movie scenes (without letting them see the actual movie scene). Ask the volunteers what type of scene they think the music accompanies. *How accurate are the volunteers? Is there overlap between types of scenes? For example, did some of your volunteers think music from a happy scene belonged in an action scene, or that music from a sad scene belonged in a scary scene?*

OBSERVATIONS AND RESULTS

Was the eerie or scary music usually in a minor key, with a relatively high-pitched melody? Did it build up tension by changing from quiet and slow to louder and faster?

There is a lot of variation in what makes up spooky music, but there are some common themes in movie soundtracks and classical pieces. The music is typically played in a minor key. For music to sound creepy or eerie, it might also be relatively quiet and slow, which can build up unsettling tension in the listener. Then at a scary or intense moment, the music may become much louder, faster, and increase in pitch, possibly with a rising scale. Higher-pitched melodies and musical instruments, such as strings and children singing (without using distinct words), are often used for these scary scenes—but other instruments, such as horns and pianos, are sometimes employed too.

CLEANUP

Put away the timer or stopwatch where you found it. Also recycle any paper that you may have used.

Discovering the Eerie Colors Behind Afterimages

HOW ARE WE DECEIVED BY GHOULISH IMAGES AROUND HALLOWEEN? LEARN MORE ABOUT ILLUSION TRICKS WITH THIS SPINE-TINGLING ACTIVITY!

Have you ever wondered how visual illusions are created? Around Halloween, we're confronted with illusions that challenge our ability to correctly perceive things, such as in haunted houses. One way in which our eyes play tricks on us is through a phenomenon called an afterimage. These are images you see after staring at an object for several seconds and then looking away. In this science activity, you'll watch afterimages to learn about how your eyes perceive color.

PROJECT TIME

15 to 20 minutes

KEY CONCEPTS

Colors
Eyes
Vision
Illusions

BACKGROUND

We perceive color using cells in the back of our eyes called cone cells. There are three different types of cone cells, and each roughly responds to red, green, or blue light. For example, when you look at a red image, the so-called red cones are stimulated and tell your brain that the object is red. The different cone cell types work together for you to see other colors, which are mixtures of these three colors. If you look at a purple image, for instance, which is a mix of red and blue, both the red and blue cones are stimulated. When all three colors are mixed, the three types of cones are all stimulated, and you see white light.

If you look at one color very long, those cone cells can become fatigued and temporarily do not respond, which is how afterimages form. As long as this lasts, you don't see with the fatigued cone cells, but you can still use your other cone cells to see other colors. After several seconds, your fatigued cones will recover; the afterimage will fade away and colors will appear normal.

MATERIALS

- Computer with a color monitor or a color printer and paper
- Stopwatch or clock that shows seconds
- Markers, colored pencils, and paper or a basic computer graphics program (optional)

PREPARATION

- To do this activity, you will need a circle that is divided into thirds (like a pie chart). The top right third should be red; the bottom third should be green; and the top left third should be blue. You can access an online version of this image here: *www.sciencebuddies.org/Files/3604/5/HumBio_img013.jpg.* For this you will need to have access to a computer with a color monitor to show the image or you can print it out on a color printer. Or if you have a circle to trace, a ruler, and colored markers, you could draw and color in the image yourself. (Try to replicate the model circle as closely as possible.) Make sure there is a white space next to your color circle that is larger than the circle.

- If there are any lights either right next to the computer monitor or colored circle printout, turn them off.

PROCEDURE

- Stare at the image of the colored circles (focusing on the small white spot in the center) for 30 seconds.
- After staring at the circle for 30 seconds, look at the white space to the right of it. *What do you see?*
- *How are the colors in each part of the afterimage different from the parts of the original colored circle?*
- **Optional:** You can use markers or colored pencils and paper or a basic computer graphics program to draw your results.
- Thinking about the rod colors (red, blue, and green) and secondary colors (yellow, purple/magenta, and cyan) and how afterimages are caused, see if you can explain your results. *Why do you think you see the afterimage colors that you do?*

EXTRA

Time how long it takes the afterimage to disappear. Then look at the colored circle for only five seconds, and again time how long it takes that afterimage to disappear. *Did it take more or less time the second time?*

EXTRA

You could try repeating this activity, but this time pay attention to how long it takes for the afterimage of each different color to disappear. *Do some colors fade away faster?*

EXTRA

Try doing this activity with several different people and have each person draw their results. *Are they all the same or are some different?*

EXTRA

You could try this activity again, but this time use objects or images that are different colors (colors other than the three primary additive ones, which were used in this activity). *Can you accurately predict what the afterimages look like?*

OBSERVATIONS AND RESULTS

In the afterimage, did you see that the top right part of the circle was cyan colored, the bottom part was purple-magenta, and the top left part was yellow?

If you stare at a red object and immediately look at a white area afterward, you will see an afterimage that is the same size and shape, but it is blue-green, or cyan, in color. This is because your eyes use the red, green, and blue cone cells to perceive white light, but because the red cone cells are fatigued, you do not see red. You are temporarily left seeing with only your green and blue cone cells.

This is the same process that happened to your eyes in this activity, and it is why the color of each piece of the circle in the afterimage is a mixture of two of the three additive primary colors (red, blue, and green), specifically the two that were not in the corresponding piece of the original image. Mixing two of the three primary colors results in the following secondary colors: red and green appears as yellow, red and blue becomes purple (including magenta), and green and blue turns to cyan.

CLEANUP

Put away the timer or stopwatch where you found it. Also recycle any paper that you may have used, and put away any markers or other materials where you found them.

Great Globs of Gluten!

LEARN THE SCIENCE BEHIND SOME OF YOUR FAVORITE THANKSGIVING SIDES WITH THIS TASTY TEST!

PROJECT TIME

45 minutes

KEY CONCEPTS

Food science
Baking
Chemistry

Do you have a favorite Thanksgiving dinner dish? Maybe it's an aunt's special cranberry sauce or mashed potatoes combined with perfectly seasoned gravy. Or perhaps you enjoy sinking your teeth into some succulent roasted turkey the most. There are also often many baked goods to choose from: dinner rolls, biscuits, cornbread, muffins, pastries, or pies! The foods in this group typically all contain a substance called gluten. In this science activity, you'll explore why some foods, all made from wheat flour, have such different levels of tenderness and toughness based on their gluten content. Then, while enjoying the bready sides on Thanksgiving, you could impress your family with your gluten knowledge!

BACKGROUND

Breads, bagels, pastas, cakes, cookies, crackers, muffins, pastries, pies, and pizza crusts. Are you hungry yet? What do all these foods have in common? Traditionally, they are all made from wheat flour, the fine powder made from grinding and sifting kernels of wheat, the edible part of that grain. Wheat belongs to the grass family, which contains some 8,000 species. Wheat was likely first cultivated in the Middle East about 10,000 to 12,000 years ago. Today, it is the third-most produced cereal grain in the world. (Corn is first, and rice is second.)

Why is wheat flour used in so many foods? When water is mixed with most other flours (such as those made from corn or rice), you get a ball of paste. Mix wheat flour with water, though, and it can make a ball that is both plastic (it can change shape) and elastic (it can return to its original shape)! Gluten, a composite of proteins within the wheat kernel, makes these valuable qualities possible. Because of these qualities, wheat dough can trap gas (such as the carbon dioxide made by yeast) and expand, which helps breads and other baked goods rise to become light and fluffy!

MATERIALS

- Two or three different flours made from wheat such as whole wheat bread flour, gluten flour, cake flour, pastry flour, or all-purpose flour (You will need 1 cup, plus a couple pinches extra, of each flour type.)
- One small mixing bowl for each flour type
- Measuring cup with 1/2 and 1/4 cup measurements
- Tap water
- Sticky notes and pen or pencil (optional)
- Fork
- Cutting board or other work surface
- Clock, watch, or timer
- Strainer with small holes

PREPARATION

- Measure out 1 cup of each type of flour into its own small mixing bowl. Keep track of which flour went into each bowl. You could use sticky notes to label the bowls. *How do the different flours look and feel?*

PROCEDURE

- Select one bowl to begin with and slowly add 1/2 to 3/4 cup (118 to 177 ml) of tap water to the bowl while carefully stirring it with a fork. It may seem difficult to stir at first but will gradually begin to come together to form a rough ball. *How does the flour feel as you stir it? How does it change over time?*

- Sprinkle a spoonful of flour (of the type you are working with) onto a work surface, and onto your hands as well, so that both are lightly dusted with flour. Place the ball of flour onto your work surface and knead it for approximately five to seven minutes until it becomes smooth and elastic. (If you are unfamiliar with kneading, this is the process of pressing down on the dough ball with your palms and then pulling it back up again and rotating it slightly with your fingertips, then repeating this over and over again.) *How does the flour change as you knead it? Does it easily form a ball over time?*

- Repeat this process (of adding water and kneading) with the other flours you are testing. Be sure to knead all of the flours for the same amount of time. At the end of the kneading process, you should have one smooth-looking elastic ball of dough made out of each type of flour you're testing. (**Note:** Some flours will form better balls than others.)

- Place your balls of dough back in their small bowls and let them rest and relax for about 10 minutes.

- Now it's time to find out how much gluten is inside those dough balls. Place a strainer in the sink where the faucet can reach it, take one of the balls of dough to the sink and cup the dough in your hands over the strainer while you let cold water run on it. Gently pull and stretch the dough apart. The water will wash away the water-soluble parts, such as the carbohydrates, but will leave behind the insoluble protein, the gluten, that developed while you were kneading. The soluble parts being washed away will look like a milky liquid. Continue rinsing your ball of dough until very little milky liquid is coming out. *How does the dough's texture and shape change as it is washed in the water?* Place the remaining ball back in its original bowl to keep track of it.

- Rinse off your strainer and repeat the rinsing process with each of the other dough balls. At the end of the rinsing process, you should have a ball of gluten for each type of flour you used! (**Note:** Depending on the type of flour you used, a lot of the ball might have been washed away.)
- Compare the sizes of the gluten balls. *Which type of flour produced the smallest diameter ball? Which produced the largest? Does this make sense based on the type of foods that each flour is used to create, and how much gluten may be in each?*

EXTRA

They won't make elegant pastries, but you can continue the investigation and bake your gluten balls by putting them in the oven at 450°F (232°C) for 15 to 30 minutes. *How do the gluten balls look, feel, and taste after they've been baked?*

SCIENCE FAIR IDEA

Repeat this using a wider variety of flours made from wheat and use a ruler to quantify your size comparisons between the gluten balls. *Which flours make the largest gluten balls (the ones with the greatest diameters) and which make the smallest? How big are the differences?*

SCIENCE FAIR IDEA

You could investigate how kneading affects the size of the gluten ball made. To do this, choose one type of flour and create three equal-size balls of dough from it using different degrees of kneading: no kneading, kneading one for five minutes, and kneading one for 10 minutes. Rinse the dough balls and evaluate the gluten balls that are created, either by measuring their diameter or weighing them. *How does the amount of kneading affect the gluten ball's size?*

EXTRA

Select one type of flour and make dough with different additives, such as salt, sugar, or oil. *How do the additives affect the amount of gluten that develops?*

OBSERVATIONS AND RESULTS

Were the gluten balls made using bread flour or gluten flour the largest whereas ones made using cake flour or pastry flour the smallest—and all-purpose flour gluten balls intermediate in size?

Typically, bread and gluten flours have a relatively high amount of gluten (around 12 to 14 percent) whereas cake and pastry flours have the lowest (about 7 to 10 percent) and all-purpose flour has an intermediate amount (around 9 to 11 percent). The more gluten a flour type has, the larger the resulting gluten ball should be. "Hard wheat" flours, such as those for bread, are high in gluten and are good for baking things that need toughness and strength such as yeasty breads, bagels, and puff pastries. "Soft wheat" flours, such as those for cakes and pastries, are low in gluten and are better for baked goods that need to be tender such as pancakes, cookies, and pastries.

Why is kneading important for measuring the gluten content in the flour? As the dough is kneaded, the chainlike molecules that make up gluten (glutenins and gliadins) link end to end to form superlong, springlike chains. This makes the tight, coiled mesh that is gluten, and this gives the dough its elasticity. All of those springs want to "spring back" when they're stretched!

CLEANUP

Wash off all of the materials you used and put them away where you found them. If you baked your gluten globs, feel free to eat them!

Salty Science
Is There Iodine in Your Salt?

LEARN THE SCIENCE SECRETS OF SALT, WHICH CAN HELP KEEP PEOPLE HEALTHY AND MAKE THANKSGIVING FOODS DELICIOUS!

PROJECT TIME

30 to 90 minutes

KEY CONCEPTS

Nutrients
Diet
Food
Health
Chemical reactions

Have you ever noticed if the salt you're using says it's "iodized"? Iodine is a micronutrient, which means we need it in small quantities to be healthy. Because iodine is relatively rare in many people's normal diets, it's added to table salt. Then when people salt their food, such as tasty turkey, stuffing, and mashed potatoes, they're also getting iodine. In this science activity, you'll use some kitchen-friendly chemistry to investigate which types of salt have iodine and which don't. Then when you sit down to your Thanksgiving dinner, you can know whether to also give thanks that you're helping combat iodine deficiency.

BACKGROUND

Micronutrients, such as iodine, are types of nutrients that people need in small amounts. Iodine is important for a person's thyroid to function normally. (The thyroid is a gland in the neck that makes key hormones.) It is found in small amounts in other foods, including saltwater fish, seaweed, shellfish, yogurt, milk, eggs, cheese, and a handful of other edibles. If a person doesn't consume enough iodine, they can become iodine deficient. The lack of this micronutrient can cause different medical problems (usually due to hypothyroidism caused by a thyroid that does not make enough hormones). These conditions include goiter (a visible swelling of the thyroid) as well as serious birth defects. In fact, iodine deficiency is the most common preventable cause of living with an intellectual disability.

Iodine (in the form of iodide) is added to table salt to help prevent iodine deficiency. Since the 1980s there have been efforts to have universal salt iodization. This has been an affordable and effective way to combat iodine deficiency around the world, but not all salt contains iodine, however. You'll investigate whether different salts have iodine by mixing them with laundry starch, which forms a blue-purple-colored chemical with iodine. (Vinegar and hydrogen peroxide are added to the salt solution to help this chemical reaction take place.)

MATERIALS

- Disposable plastic cups that are 10 ounces (296 ml) in size or larger. (Alternatively, you could use smaller cups and scale down the activity.)
- Distilled water
- Measuring cups
- Measuring spoons
- Laundry starch solution, also called liquid starch (Alternatively, you could make a suitable starch solution by dissolving 1 cup [237 ml] of starch-based biodegradable packing peanuts in 2 cups [473 ml] water.)
- Iodine antiseptic solution (optional) (Use either an iodine tincture or povidone-iodine solution, found in the first aid section of grocery stores and drugstores. If the iodine doesn't come with a dropper, you'll also need a medicine dropper.)
- Disposable plastic spoons

- At least three different types of salt to test—for example, plain (noniodized) table salt, iodized table salt, pickling salt, rock salt, kosher salt, "lite" salt, and sea salt (If you aren't using the iodine antiseptic solution, include iodized table salt.)
- 3 percent hydrogen peroxide
- White vinegar

PREPARATION

- If you are using an iodine antiseptic solution, you can prepare a positive control cup so you know what the reaction between iodine and starch should look like. Pour 1/2 cup (118 ml) of distilled water into a disposable cup, add 1/2 teaspoon (2.5 ml) of laundry starch solution, and then add five drops of the iodine antiseptic solution. Be careful when handling the iodine because it can stain.
- Stir well with a disposable plastic spoon. *What happens to the liquid when the iodine is added?*

PROCEDURE

- Pick one of the types of salt you want to test and measure out 4 tablespoons into a clean, plastic, disposable cup. Add 1 cup (237 ml) of distilled water to the salt and stir well for about a minute with a clean, disposable plastic spoon. You do not need all of the salt to dissolve.
- Then add 1 tablespoon (15 ml) of white vinegar, 1 tablespoon (15 ml) of hydrogen peroxide, and 1/2 teaspoon (2.5 ml) of starch solution. *What do you think the purpose of the starch is?*

- Stir the salt solution well with the disposable plastic spoon and then let the solution stand for a few minutes. *What happens to the solution after you stir it? Does it become a blue-purple color?*
- Repeat this process using the other, different types of salt you want to test. For each type, be sure to use a different, clean disposable cup and spoon. *Do any of the other salt solutions become a blue-purple color?*
- *Based on your results, which salts do you think contain iodine (in the form of iodide) and which do not? Do your results agree with the labeling on the salt packages, which often say whether the salt contains iodide or not?*

EXTRA

Try this activity with even more different types of salts. For some ideas, see the Materials list. *Which types of salt contain iodine and which do not? Do your results agree with their labeling?*

EXTRA

In this activity, you added vinegar because it is an acid and helps the chemical reaction take place. Try testing the iodized salt solution again, but this time leave out the vinegar. *Does the reaction still take place, turning the solution a blue-purple color? If the reaction occurred, did it take a longer amount of time to happen?*

EXTRA

Temperature often affects chemical reactions. You could try this activity again, but test an iodized salt solution at different temperatures (by heating or cooling the distilled water). *How does changing the temperature of the solution change how the color-changing reaction takes place?*

OBSERVATIONS AND RESULTS

Did the iodized table salt solution change to a blue-purple color when you mixed in the starch? Did the "lite" table salt similarly change color whereas most of the other salt types did not?

In this activity, you should have seen that the iodized table salt and the "lite" table salt solutions both changed to a blue-purple color (as did the iodine antiseptic solution, if you used it). This indicates that iodide is present in these types of salts. You likely saw no color change for the solutions made using the noniodized salt, rock salt, kosher salt, or sea salt because these varieties do not typically contain iodide.

The starch solution was used in this activity because it forms a blue-purple chemical when combined with iodine. Because the solution's original pH needs to be changed for this chemical reaction to effectively take place, vinegar (an acid) is also added. Hydrogen peroxide is used to turn the salt's iodide into iodine, which the starch reacts with.

CLEANUP

Be sure to thoroughly wash any measuring spoons or other utensils that came into contact with the solutions made in this science activity. You can dispose of the solutions by pouring them down the drain.

THE SCIENTIFIC METHOD

The scientific method helps scientists—and students—gather facts to prove whether an idea is true. Using this method, scientists come up with ideas and then test those ideas by observing facts and drawing conclusions. You can use the scientific method to develop and test your own ideas!

Question: What do you want to learn? What problem needs to be solved? Be as specific as possible.

Research: Learn more about your topic and refine your question.

Hypothesis: Form an educated guess about what you think will answer your question. This allows you to make a prediction you can test.

Experiment: Create a test to learn if your hypothesis is correct. Limit the number of variables, or elements of the experiment that could change.

Analysis: Record your observations about the progress and results of your experiment. Then analyze your data to understand what it means.

Conclusion: Review all your data. Did the results of the experiment match the prediction? If so, your hypothesis was correct. If not, your hypothesis may need to be changed.

GLOSSARY

antiseptic: A substance that prevents or destroys the growth of microorganisms such as bacteria often in or on living tissue, such as the skin.
biodegradable: A description of substances which decay naturally without damaging the environment.
carbohydrate: A substance that gives your body energy and is found in foods including sugar and potatoes.
chemical reaction: A process which involves change to the structure of atoms or molecules that make up a substance.
compound: A chemical which is a combination of two or more elements.
deficient: Not having enough of something, such as a vitamin.
exothermic: Describing a reaction that gives off energy in the form of light or heat.
hydrogen peroxide: A liquid that is clear and can kill bacteria, remove color from cloth, and can bleach hair.
hydrophilic: A word to describe substances which can be mixed with or dissolved in water.
hydrophobic: A word describing substances that cannot be mixed with or dissolved in water.
hypothyroidism: A condition that involves the thyroid gland not making enough thyroid hormone which can cause problems with the heart, nerve function in the hands and feet, and mental health problems.
iodization: The process of iodizing, or treating something with iodine or an iodide.
molecule: The smallest part of a substance that has all of the properties of that substance.
pH: A measure of a solution's acidity and alkalinity that is a number on a scale of up to 14. The number 7 represents neutrality and higher numbers on the scale represent increasing alkalinity. Lower numbers represent increasing acidity.
salmonella: A type of bacteria that can be found in food that is not cooked enough and can make people very sick.
soluble: Able to be dissolved in or as if in a liquid, often specifically water.
tincture: A solution which includes a medicinal substance in an alcoholic solvent, or liquid able to dissolve or disperse another substance.

ADDITIONAL RESOURCES

Books

Dickinson, Michelle. *Kitchen Cabinet Science Projects*. New York, NY: Penguin Workshop, 2021.

Maclaine, James, Lizzie Cope, Rachel Firth, and Darran Stobbart. *Science Experiments*. London, UK: Usborne Publishing Ltd, 2022.

Yi, Andrea Scalzo. *Super Simple Science Experiments for Curious Kids*. Salem, MA: Page Street Publishing Co., 2022.

Websites

Discovery Education
sciencefaircentral.com

Exploratorium
www.exploratorium.edu/explore/activities

Science Buddies
www.sciencebuddies.org/science-fair-projects/project-ideas/list

Science Fun for Everyone!
www.sciencefun.org/?s=science+fair

Videos

Exploring the Science of Sound at Home | Science of Sound
ny.pbslearningmedia.org/resource/exploring-the-science-of-sound-at-home-video/science-of-sound/

Five Physics Experiments for the Holidays | Physics Girl
ny.pbslearningmedia.org/resource/holiday-physics-physics-girl/holiday-physics-physics-girl/

Kitchen Science | DIY Science Time
ny.pbslearningmedia.org/resource/b4ad5b8d-d11f-4bde-8aac-2254648fbf1a/kitchen-science-video-diy-science-time/

Publisher's note to educators and parents: Our editors have carefully reviewed these websites to ensure that they are suitable for students. Many websites change frequently, however, and we cannot guarantee that a site's future contents will continue to meet our high standards of quality and educational value. Be advised that students should be closely supervised whenever they access the internet.

INDEX

A
acids, 6, 23, 24, 27, 30, 58, 59
adsorption, 24
amplification, 11

B
baking soda, 6, 29, 30, 31, 33, 34
bread, 6, 9, 49, 50, 53

C
carbon dioxide, 6, 9, 30, 34, 50
chemistry, 5, 17, 29, 49, 55
combustion, 30

E
emotion, 39, 40, 42
exercise, 11, 12, 13, 14, 15
eyes, 33, 37, 45, 46, 48

F
Fourth of July, 4, 23, 29

G
gravity, 29, 30

H
Halloween, 4, 35, 39, 40, 41, 45
heart, the, 11, 12, 13, 14, 15

K
kneading, 51, 52, 53

L
liquids, 6, 8, 17, 18, 19, 20, 21, 24, 51, 56, 57

M
micronutrients, 55, 56
music, 39, 40, 41, 42, 43

N
New Year's Day, 5

P
photoreceptor, 35, 36
pressure, 29, 30, 34

S
science fair ideas, 8, 9, 14, 26, 33, 42, 52
soap, 17, 18, 19, 20, 21
St. Patrick's Day, 17
surface tension, 17, 18, 21

T
Thanksgiving, 49, 55
thrust, 29, 30, 34
thyroid, 56

V
Valentine's Day, 11